A History of Time

humbly submitted

by

Paul Janson

JM Publishing

Published by JM Publishing: July 2, 2019

Author Photograph: Sarah McGrath

All rights reserved

For more information contact:

Paul Janson
JM Publishing
64 Elm Street
Georgetown, MA 01833

Email   Paul_janson@AOL.com

Web Site   PaulJanson.com

Blog:
https://pauljanson.wordpress.com/category/philosophy/physics/

ISBN 13: 9781079568912

Copyright 2019 Paul Janson

A History of Time

First let me offer an apology to Steven Hawking for borrowing and modifying his title. My effort will be to briefly review cosmic time only as a background and then to spend the bulk of my effort on the history of artificial "time", the time that we as humans have imposed on the space between the events of our lives: the years, days, minutes etc. by which we divide and measure our lives. The real, physical time, the kind Dr. Hawking dealt with, is not the overriding subject of this writing, although as we shall see, separating the two is in truth impossible.

Cosmic Time

This term is one which I prefer as it distinguishes time in the rest of the universe from time in our small part of that universe, in our lives, on our planet and important only to us. Time in this small morsel is what really matters just as the weather in your city will determine how your ride to work or to the grocery store will go.

To begin this abbreviated discussion, it is necessary to state that time did not exist in the beginning. The current theory regarding the formation of the universe is the Big Bang Theory, not to be confused with the TV sitcom which is in fact much better recognized and more popular than anything I will write here. You may read my thoughts on this subject on my WordPress blog.

According to Big Bang Theory (there are yet some difficulties with it) time began when the Bang occurred, not before. One could not wait for the Bang since there was no time. This is of course not a theory of creation since one of the premises used to form the theory is that everything existed at the initial Bang; it was just in a different form than it is now. It was all energy gathered together into a single point: no space, no matter, and no time: an interesting situation. The major support for the theory is that the universe is expanding, and therefore yesterday it must have been smaller. Last year: smaller still and so on. Eventually one can go back to where the universe could not get any smaller, a single point. That calculates out to fourteen billion years, approximately.

But this leads to two interesting conclusions. The first is that events we see on distant stars or galaxies have produced light many years ago for us to see. Some events that we are just seeing now have traveled so far that they actually occurred when the universe was quite young, shortly after the Big Bang. When we look at the night sky, without the bright sunlight to obscure the less intense stars, we are in fact looking at the universe over an enormous period of time – all at once. The events that we see on a star very close to Earth may have happened only a few thousand years ago, while its neighbor may be a very distant star whose events happened billions of years ago. The latter star may very well no longer exist certainly not in the form or shape that we see it. A reminder here: the closest stars to Earth are in the Alpha Centauri complex. The two main stars form a binary pair, Alpha Centauri A and Alpha Centauri B, and are an average of 4.3 light-years from Earth. The third star in this complex is Proxima Centauri, about 4.22 light-years from Earth and it is the closest star to us other than the sun of course, hence its name: _Proxima_ Centauri. Another reminder is the terminology used here. A light year is the distance light travels in a year (5.879 with 12 zeros miles or 9,461 with 12 zeros again in kilometers). For this example, it is the time that is noteworthy. Events we see today when we look at Alpha Centauri occurred ONLY 4 years ago. That's before I began collecting Social Security.

To put perspective on this, events taking place on the moon, like Neil Armstrong taking, "one small step for man", occurred 1.3 seconds before their transmission arrived on Earth, both the video he sent and his voice. He stepped onto the lunar surface 1.3 seconds before we saw it. NASA was actually concerned that the delay would confuse people. It turned out to be a non-issue.

Events on the Sun occur approximately 8 minutes and 20 seconds before they are seen on Earth. That bothered astrophysicists enough to prompt them to develop the theory of a curved fabric for the universe to explain the immediate far-reaching effects of gravity. It also explains why gravity, which pulls on things with mass, can bend a light beam even though light, or photons, the particle form of light, has no mass. The photon follows the curved fabric of the universe, the curve being the result of the mass of objects in the universe.

The second interesting conclusion relates to Einstein's work a century ago, before anyone collected Social Security. I will present the conclusions only, because the explanation is rather involved and because I'm not sure I understand it myself. It relates to the interaction of matter and everything else in the universe, including time. What Einstein said is that matter and energy is the same thing, or more precisely they are interchangeable. They can be converted from one form to the other using his famous formula $E = mc^2$ or energy is equal to the mass of matter multiplied by the speed of light squared (a very, very big number), a little less than 300 million meters per second squared, that is times itself (approximately 9 with 16 zeros). When one takes a small amount of matter and converts it into energy, the amount of energy is huge: an atomic bomb huge.

Also predicted by quantum physics is the fact that as an object travels faster, time on the object slows down. At the speed of light, time stops. The object's mass also increases as the object's speed increases, so traveling at the speed of light would result in an object having infinite mass and no time. Photons can travel at the speed of light because they have no mass, and of course they do travel that fast because they ARE light. Gravity will also slow time so a black hole will have very slow time. These are the logical conclusions of Einstein and other physicists' work and appear confusing because they are confusing. I present this for completeness sake and to prepare you for the answers to simple questions such as: Why are there seven days in a week and twenty-four hours in a day? The answers here are much less confusing as you will see if you continue reading.

But where does time and space come from anyway? They didn't exist prior to the Big Bang so they must have been created after the Bang, but since the universe is expanding and time is marching forward, they must be being produced now as well. What are they created from? There is a theory, which I believe has merit, that time and space are created out of energy and matter. That is all four: energy, matter, space and time are in fact interchangeable. If this is true then it follows that the universe is finite since eventually everything will be converted into time. That is: all of the other three: matter, energy and space will become time and the universe will end – when it runs out of things to make into time. Just a thought, no proof offered yet anyway.

One final point: time travel. To travel backwards in time, one would need to travel faster than the speed of light. I have offered the standard explanation for why this is impossible. At the speed of light mass would become infinite and therefore to increase speed even a little would require an infinite force be applied, and no such force exists. Now let me further explain what would really happen if you could. First time would move backwards, but only for the things traveling faster than light. If you did travel faster than light, you would grow younger and your vehicle, which is also traveling faster than light, would go backwards in time. That is it would grow younger, and depending on how far back it went, it would begin to disassemble, becoming the parts from which it was made and eventually the raw material from which the parts were made, and so on. When you landed in your hometown or wherever, you and the vehicle would be younger having traveled back in time, but (this is important for all the sci-fi people out there), the town would have traveled at its usual time and be older. For the town to travel backwards in time, you would have to take the town in your time traveling vehicle. To see a dinosaur, you would have to bring its remains with you in your vehicle, its complete remains, a few bones would not suffice since only what travels faster than light will go backwards in time. Incidentally, light does not get "old" since it always travels at the speed of "light" and the ancient events on distant stars did occur years ago since the stars are not traveling at the speed of light, only the light from the stars. Time passes normally on those stars.

Now for the simple stuff.

Time dominates art

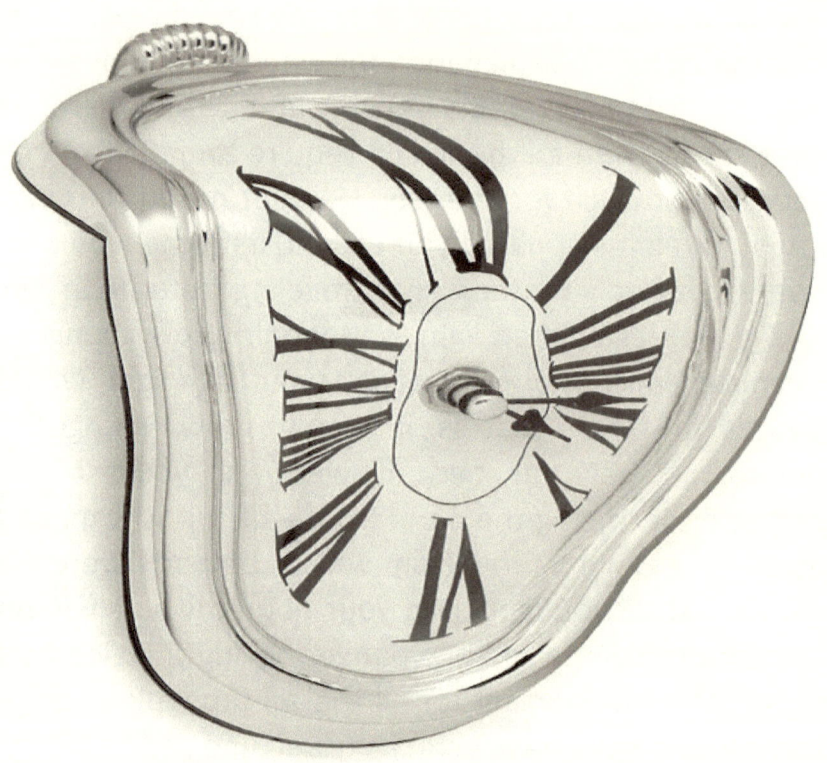

A portion of the painting by Salvador Domingo Felipe Jacinto Dalí i Domènech, 1st Marquis of Dalí de Púbol, commonly known simply as **Salvador Dali**. The painting is titled:

<u>**The Persistence of Memory (La Persistencia de la Memoria)**</u>

It was created in 1931 and displayed in the Museum of Modern Art in New York City since 1934. A minor note: the number four appears to be the Roman IV. Most clocks use the Roman IIII which is more balanced with the eight, VIII, opposite it on the dial.

Our Time on Earth

I will begin this section by reminding the reader that the previous discussion was about universal time, determined by the physical properties of the universe. The discussion from this point forward is about the division and measurement of time that is partially or completely artificial. The year is defined by the Earth's orbit around the Sun and its own rotation, but the divisions of that time are defined by people. The length of an hour, or the number of hours in a day follow the rules people have made for them, not the rules that govern the universe.

The events that we take as certain and consistent are not at all as they appear. The day is determined by the rotation of the Earth on its axis, as I have said, but that day, the time for a revolution adjusted for the travel around the sun, has not always been the amount we experience today. The Earth is spinning as would a top, a very large top, but the spin is gradually slowing as it is with all planets and moons and so on. That is to say, millions of years ago, the day was shorter: i.e. the Earth was rotating faster. In fact the year must be adjusted by a millisecond of time now every few years because of this slowing of the Earth's rotation. The natural period of rotation for a heavenly body is one that will be equal to the period of revolution a body has around its central gravitational object, the Sun in the case of the Earth and the other planets. That is where the Earth will eventually settle if the solar system lasts long enough. The planet Mercury has reached this point, probably because it is closer to the Sun and relatively small, and the Earth's moon has done the same, possibly for the same reason.

Incidentally the moon was much closer to the Earth at the time of its birth which occurred between 20 and 100 million years after the solar system began forming, about 4.5 billion years ago. A Mars size object hit the Earth in the Theia Impact or what is called The Big Splash. The moon formed from the debris left by the impact. Our moon is the largest natural satellite or moon, relative to the size of the planet it orbits. It was much closer to Earth as I said and in fact is still moving away from Earth at about an inch every year. Nothing stays the same.

    The other planets would indicate that size may be the major determinant. Venus, the second smallest planet after Mercury now that Pluto has lost its standing, takes 116 Earth days per rotation and approximately twice as long per revolution, 225 Earth days. Jupiter, much larger and further from the sun than Earth, rotates on its axis in 9 hours 56 minutes, but revolves around the sun every 12 Earth years. Mars is the planet between Earth and Jupiter and is the closest in size to Earth. Its rotation is approximately the same as Earth, while its revolution around the sun, its year, is almost twice that of Earth, *687 Earth days*. Uranus is the planet beyond Saturn, and is smaller than Jupiter and Saturn, but it is more massive than Neptune the next planet. Uranus is denser and therefore smaller in size but has a larger mass. It has a rotation of 16 Earth hours and a revolution of 84 Earth years, while Neptune rotates on its axis in 17 Earth hours. Its journey around the Sun takes 165 Earth years which calculates out to approximately 233 Neptune days for a revolution around the sun. Neptune was only discovered in 1846 so it completed its first orbit around the Sun since its discovery, 11 Earth years ago, in 2011. All of this depends on the speed of rotation when the planet was formed of course. Incidentally the light from Neptune takes four hours to reach Earth. Pluto, the planet, was discovered in 1930 and demoted to dwarf status in 2013. Its trip around the Sun takes 248 Earth years, so poor "planet" Pluto didn't even get to complete its first orbit (since being discovered that is). Does that make Pluto the only "planet" that never completed an orbit around the Sun? It was a planet for a few (Earth) years. For a further look at my thoughts in this direction see my article: **The Universe and Time**, on my WordPress blog.

Now for the measuring of time. This is the real subject of this writing: not what time is, but rather how we measure it. The problem faced by our forbearers was that there was obvious order to the universe and it was from the outset assumed that the Earthly events were connected to the universe in a logical way. It was also assumed that it was the heavenly events that determined the Earthly events or at least influenced them. This persists to our day in subtle ways; for instance, the disease influenza was named because it was a seasonal occurrence therefore assumed to occur because of the *influence* of the stars. This is the basis for astrology and zodiac signs and all of that. It is not really that surprising to believe this except that we do not observe the skies as our predecessors did. We don't notice that the tides follow the moon, or that the full moon is always rising with the setting sun. This latter connection must be so since the moon can only be full if the sun is opposite to it to shine its light directly on it. That logic was clear to our ancestors, but they had an advantage: their night sky was not "polluted" by artificial light from neon, LED's and all the other sources that light up our nights. Then again they could not safely venture out at night as we routinely do.

Let us start with the year. This seems simple, but it is not. The word "year" is derived from old German or Anglo-Saxon *jer* meaning season. The belief that the universe was an orderly, connected thing led to the belief that the Sun's motion controlled Earthly events, and time had to be orderly. The year had to be a number that was orderly and predictable. Now today we all know that the year is 365 days long except for the leap year, every fourth year, when it is 366 days long unless ... well more on this later. What we think we know is of course wrong. The year is not 365.25 days long as it would be if a full day were added every fourth year, but 365.2422 days or 365 days, 5 hours, 48 minutes, and 46 seconds – approximately. There are other "years" such as the *solar* or *tropical year* based on the sun's position, and the *northward equinox year* which attempts to place the vernal equinox, the first day of spring in the northern hemisphere, on or near March 21st. There is also a calendar that attempts to have the new year occur exactly on the vernal equinox called the Solar Hijri Calendar. This was formulated in 1925 as an update of the historical Jalahi calendar. Incidentally it only accomplishes its purpose in Tehran, Iran. None of these considers the gradual slowing of the Earth's rotation and consequent lengthening of the day.

But the Earth is slowing its rotation as I said above and the dinosaurs, say 400 million years ago enjoyed a shorter day perhaps only 20 or 21 hours long. There are several attempts to calculate the exact number of hours, but all make assumptions and estimations along the way, so they are not worth exploring in detail. The principle of shorter days is important, however, at least to the dinosaurs. Consider what this would mean: less diurnal temperature variation. That would make temperature regulation less of an advantage. It is temperature regulation that makes mammals more adaptable than dinos. The current theory about a meteorite or comet disrupting the Earth's atmosphere begs the question: if the dinosaurs could evolve and adapt before the meteorite, and beat out the mammals, why didn't they do the same after the impact? If you are a Darwinian believer, the answer is that they were less suited, did not *fit* the environment any longer. They hadn't changed but the environment had. Not some short-term change caused by dust in the atmosphere for a hundred or even a thousand years, but a longer day and greater temperature variation from day to night might explain the phenomenon better. An impact may have tipped things, but the dinosaurs couldn't adapt to a changed world – unless they grew feathers to help keep them warm at night.

    I have digressed however and there is ample excitement in the human measure of time. The makers of the first calendars tried to make things come out to be an even number. Their expert observations proved that that was impossible. They really wanted the year to be 360 days long, a nice, neat, number, divisible by so many other numbers. This is evident in other measures: the 360 degrees in a circle for example. The year was so close they couldn't resist. Many early attempts had a 360-day year with 5 days for religious festivals somewhere along the way. They were truly expert observers though and that forced them to adopt the 365-day year with a day added every fourth year. Julius Caesar brought this back from Egypt to Rome along with Cleopatra.

This was called the Julian calendar. The calendar it replaced was the Consular Calendar which began on the date the consular took office. The Romans were in the habit of changing the number of days in a year (calendar), in order to stay in office longer. What's wrong with a 400-day year? Julius put an end to that, basically by saying he would be Emperor for as many years as he wanted so no need to lengthen the year. He did want a month named for himself though. Now the Julian calendar at the time started in March; the Vernal equinox was the beginning of a new year, and the months were named for deities: Janus got January, Februus is believed to give his name to February although some sources maintain it is named for the Roman festival of purification, Februa. Mars got March, and so on through the month of June. The remaining months after June were numbered. We still keep evidence that this is true since September, October, November and December are numbered months: seventh, eighth, ninth and tenth. The eleventh and twelfth month were January and February. There were five months with 31 days, and the rest had 30 days adding up to 365 day total. Julius chose the month after June, named for Juno, Zeus's wife, and mother of six of his twelve children to become his month. He couldn't very well steal Juno's month. The only problem was that the month he chose and named July for Julius, had only 30 days. That could not be allowed to continue, so he simply took a day from February, the last month of the Roman year, and added it to his month. Easy enough.

Now it turned out, of course, that Julius did not reign as long as he wanted, but he did get to keep his month. He was succeeded by his nephew, Octavius, who took the name Augustus and should probably be credited with building the Roman Empire. He added a little territory during his 44-year reign, not as much as Julius had, but he did establish a bureaucracy that survived some pretty bad emperors. For this he deserved a month, and he got the one following Julius, just as he followed Julius as emperor. A minor problem was that his month had only 30 days too, but Julius had shown the way to solve that problem. August ended up with 31 days while February diminished to the shortest month with only 28 days. It did get the leap year day as a consolation. February was the last full month of the Roman calendar and was therefore the month during which debts were paid and disputes settled in order to begin the new year right. Losing a day or two of that was welcome perhaps.

We are almost done with the year so hang in there. The trouble with the Julian arrangement is that the year is not really 365.25 days long, but 365.2422 days long as I mentioned above, so the calendar slipped forward a little with each passing year. This eventually became a problem, a millennium and a half after Julius and Brutus had it out. The calendar was no longer functional for things like farming, planting, or religious holidays. In October of 1582, Pope Gregory XIII proposed a change to bring things back in line, the Gregorian calendar. By this system the leap year occurs every year divisible by four, unless it is divisible by one hundred in which case it is not a leap year, unless it is divisible by four hundred in which case it is a leap year. That is why the year 2000 was a leap year. Oh yes and the calendar had to be reset to make the holidays fall where they should. October was the month from which the days, 10 in all, were subtracted, but only that one time. Of course, landlords and holders of loans insisted that they should be paid for a full month even though the days were missing. The Gregorian year is 365.2425 days long which is close enough. I have seen mentioned a system that introduces exceptions to the leap year in the years divisible by one thousand (a leap year), and four thousand (not a leap year), but this is not used to make the current calendar.

The reformation had occurred before Pope Gregory proposed his calendar so of course the Protestants refused to adopt Gregory's new calendar for some time. This led to minor oddities such as the Glorious Revolution of 1688 occurring in 1689 by Gregory's reckoning. The new year had been moved to January probably to be closer to Christmas. This is all referencing the standard or solar calendar. The Asian countries, Muslims, Jewish and Eastern Orthodoxy still calculate their year using the lunar cycle. This causes the holydays to move around excessively, and so Gregory's calendar became standard for world finance and other communications. Asian countries had accurate calendars, but not uniformly consistent, changing with ruling parties. These countries have adopted Gregory's too for daily life and business. The last Eastern country to adopt the Gregorian calendar did so in 1923.

If the year and its months seem to overwhelm, then the seven-day week will give a respite. Why seven? This goes back to the celestial-Earth connection again and it is obvious if one takes the names given in a Romance language like Spanish: Lunes, Martes, Miércoles, Jueves, Viernes, Sabado and Domingo: or: Moon day, Mars day, Mercury day, Jupiter day, Venus day, Saturn day and God's day or the Sun's day. Uranus, Neptune, and Pluto had yet to be discovered. They would get their chance as the larger atomic elements were discovered or created: uranium, neptunium, and plutonium. One has to hope that the anti-Pluto groupies won't demand that plutonium change its name too or that it be demoted to a "dwarf element".

Now for the explanation for the names of the days of the week. The heavenly firmament that governs the Earthly events is not really "firm" at all. When viewed with unassisted eyes, no telescopes, there is an unchanging pattern of stars with seven heavenly bodies that sort of wander aimlessly about. The Sun, the Moon and the five visible planets and each of these was given control of a day of the week. These wandering stars (the Sun, which is the only real star, the Moon, Mercury, Venus, Mars, Jupiter, and Saturn), play a great role in astrology even today and can be consulted in the back pages of newspapers next to the crossword puzzles. This was even more important back in the olden days. Johannes Kepler, the astronomer who showed that the heavenly bodies moved in elliptical not circular orbits, was required to draw up astrological charts for his boss, Emperor Rudolf II, as a condition of his employment.

The English names for the days of the week are a mixture of the Latin and Germanic, but the pattern persists. This is evidence of how strong the belief in the heavenly influence on Earthly events truly was. All other measures of time were dominated by the belief that they must fit "logical" numbers, easily divided into groups. The days of the week stand out in sharp contrast here. An aside here: during the French Revolution there was an attempt to establish a ten-day week, with a three-day weekend, but this was uniformly rejected.

It may be worth digressing here to mention the extraordinary power of numbers in the period we are exploring. Mathematics held a near religious position. I have read that the Pythagorean school actually executed people for daring to discuss irrational numbers, such as the square root of two in spite of the fact that the Pythagorean theorem predicted its existence. This was at a time when science and religion were different parts of a single study and what we call science today was referred to as natural philosophy. Science and religion were not the opposing forces that we see today, just a different explanation of the universe.

Now, however, it is time to move on to a discussion of hours, or more accurately the divisions of the day. The word *hour* derives from Latin: *hour*, Greek *hora* or French *hore* meaning one 12th of a day. The root is also carried over into *horo*scope and *horo*logist (a studier of clocks). Numbers again dominate here. The day was divided into 12 equal hours in order that it would be easily divided into pieces: just the day however, not the night. This seems to have begun when the sundial was introduced in Rome. It was "introduced" by being looted from the Greeks in Sicily. These sundials were designed for a slightly lower latitude and were therefore inaccurate by a small amount, but this was tolerated. There were public sundials, or solariums, one of which was a large device at the Roman forum. Many of the wealthy and important Romans had sundials in their private gardens and there were even portable sundials devised. I have a replica of one of these and can attest to the fact that it functions fairly well if one knows how to use it.

As I have said, the measurement of time at this point did not include the night. The day began with sunrise and the first hour was followed by 11 more equal hours for a total of 12. If for instance, one reads the Biblical account of the crucifixion of Jesus, we see that he died in the ninth hour, but the sun had not yet set. There was still time to take him down from the cross and bury him before sunset and the beginning of the Sabbath. The ninth hour occurs at approximately three o'clock in the afternoon, or more precisely three hours before sunset since there were always exactly 12 hours of daylight in the day. As with many of the conventions present at the time of the formation and division of our time, the language persists even today. We commonly speak of the "time of day", but seldom of the time of night. Two thousand years ago there was no time at night or at least no measured time since there was no sun to measure it by. Before we start criticizing the ancients for having variable hours filling the day, consider that if the sun, and the sundial determined the hour of the day then it makes a great deal of sense to have the same number of hours with a varied length of the hours, then to have a fixed hourly length and vary the number of hours in the day. The latter situation would, of course, require fractions of an hour to be used. When the day and night were combined the time could be fixed at 24 hours and a fixed hourly length made sense.

Two other conventions should be mentioned here before moving on to the conquest of night. First is the use of the term o'clock. This refers to time measured by a clock (Of clock, or of the clock). This is to distinguish it from time measured by the sun, as was once the only way to measure time during the day. Sun dials were the standard and time was not measured at night. The development of the clock and its impact will be discussed later.

The second convention regards the moon's influence on human activity, not the astrological influence, but the very practical influence. As science and learning advanced during the era called the enlightenment, roughly 1600 to 1800 and also often referred to as the *Age of Reason*, groups of scholars began gathering regularly in the evening to discuss various social and scientific ideas. These gatherings were often referred to as a *Lunar Society*, most notably in Birmingham, England. It was held on the evening of the full moon each month because the full moon rose with the setting sun and set with the rising sun, as discussed above, and thus providing good light for gentlemen to be wondering about at night. Tall case clocks, sometimes called grandfather clocks, often have a dial above the face that shows the phases of the moon.

After appreciating these incidentals, it is time to return to the hours of the day – and night. The Romans conducted their business during daylight, but the real stimulus for the standardization of the hours was the requirement of the monasteries. Saint Benedict of Nursia in 525 AD was among the first to set specific hours for prayers, the canonical hours: these are matins and lauds, prime, terce, sext, nones, vespers and finally compline. They correspond to dawn or first light: martins and lauds, prime - the second canonical hour occurs at sun rise, tierce the third hour, sext the sixth, nones the ninth, vespers at sun set and compline at dusk when the final day light disappears. By our modern time of day, the approximate times are: martins and lauds are at 3 or 4: a.m., prime is about 6: a.m. terce the third canonical hour at about 9: a.m., sext the fourth hour at noon, nones the fifth of the seven canonical hours; about 3: p.m., vespers at 6: p.m. and compline at 8 or 9: P.M., during summer months that is. The day obviously still had 12 hours beginning at dawn with the first hour as is evident if the names given to the canonical hours: prime (one) at sunrise, terce at the third hour and so on. The day was shorter, the day light that is, in winter and longer in summer so the monks slept less in summer but made up for it in the winter. Sort of a *daylight savings time* in an early time.

This idea was borrowed from Jewish practices and eventually handed to Muslims. The Muslim prays five times a day at prescribed hours in a prescribed fashion referred to as Salal. The above system is usually referred to as *real solar time* determined by the sun's position. This is in contrast to *mean solar time* which has the 12 or 24 hours of constant length with the sun rising and setting at differing hours of the day but with noon occurring at the local zenith of the sun in each locale.

One further myth surrounding monostatic life; a great deal of time was spent praying of course, but time was also spent copying manuscripts. The word "manuscript" means handwritten. These activities were all governed by the daily canonical hours. Monks are often pictured copying these manuscripts laboriously by candlelight, but that is fiction. First the candle was expensive and provided poor light to copy by.

Second and perhaps more important, sitting sleepily in a cell with valuable and very flammable paper books was dangerous, not only for the monk but for the books. Pray at night and copy in the daylight. A great deal is made of the beautifully illustrated books these fellows produced, the illuminated texts. Why so? Part of it was the artistic nature coming out of the monks drab robe and drab life, but it's also true that the vast majority of humans at the time could not read, thus the demand for the books was not that great. There was no sense in producing large numbers of a product with a small literate market.

At this point it seems appropriate to present the advent of the clock, and while it evolved over centuries and in fact is still evolving, let us begin here. The first step was to appreciate that it was needed. When time was only important during daylight, the problems were relatively small. Mostly these problems involved clouds or other weather phenomenon that prevented the sundial from functioning. These difficulties affected few people and even those affected were only slightly inconvenienced. Daily activities were not much governed by precise hours. There were some, however, for instance the trial system. An orator was allotted a specific length of time to defend or accuse the client. Usually this was an hour or so and the Romans managed to succeed in enforcing this restriction. The time was measured by an hourglass or a water glass. This is actually a simple device to manufacture and to calibrate. A vessel was fashioned with a small opening through which sand or water could slowly pass. In the case of an hourglass, the sand passed through into a lower chamber so that when inverted it could run back. The size of the passage was not critical since the time, an hour usually, was measured against a known, accurate hourglass. The only limits on the hole were that the sand couldn't be trapped in it (too small) or run out too quickly (too large) but otherwise a large range of size would work very nicely. When the time passed, an hour say, the amount of sand was set and what remained was poured out leaving just enough to pass through the hourglass in an hour's time. The sand was matched to the passage, not the passage to the sand. So simple and so successful was it that it was used long after accurate clocks became available. For water a similar procedure was followed except that the water simply passed through the hole and at the end of an hour the remaining water was again discarded and the bowl into which the water would be placed was marked at the appropriate level using the water that had passed in the hour. A note here: this hour was of a constant, not a variable length. This shift was the beginning of the standard hour.

The principal use of the hour glass or water glass was to time orations, the length of time a person could speak particularly during legal orations, i.e. trials or debates. While this restriction was abandoned, unfortunately, the language it inspired lingers. We still use the phrase: "Time is running out", a reference to sand or water running out of its vessel thus silencing the orator or lawyer. In Puritan New England the hourglass was used to time the minister's sermon. He was required to preach for a minimum time, usually three hours. The congregation felt they deserved that much for supporting him.

I'll point out now, if the reader has not already noted, that our language today has imbedded in it references to the historic origins of the measurement of time, as well as other commonplace things around us. If you're wondering where the clock is in all this, let me say that the hourglass was in fact one of the earliest attempts to make a clock, a machine to measure time.

The clock as we know it would not appear for some time, but the idea was there, and the need was obvious. The circular "dial" on the clock was modeled after the "dial" on the sundial and rotated "clockwise" because that is how the sundial worked. The Sun moved from east to west, so its shadow moved from West to East or clockwise. The moving of the first hour to the middle of the day (and the top of the clock dial) was a matter of convenience. The Sun's zenith was easily determined, and the clock could be easily corrected each noon hour for any error that had occurred. All time at this point was strictly local time. There were no time zones yet.

Darkness by cloud or night made solar time impractical. Monasteries were probably the first place to solve the problem: monks had a very structured life what with prayers and copying and all the rest. The word *clock* is derived from Medieval Latin, Dutch or French (clocca, clocke or cloke) a word meaning "bell", because these early clocks announced the time by the strike of a bell. The original monastery clocks were adjustable to accommodate the varying length of the hours. The real problem was regulating the machine.

Early versions were adjusted to swinging balls and other such devices, but it was the pendulum, thank you Galilei Galileo, that made the clock accurate. Galileo is supposed to have realized that pendulums swung at precise periods depending on their length, while watching chandeliers swinging in church. Apparently Galileo did not feel the priest had anything that important to say.

The pendulum was coupled with the anchor escapement which regulated the movement: the hands of the clock. The anchor escapement has its name because it kind of resembles a ship's anchor. The escapement is attached to the pendulum and with each swing allows a saw-toothed wheel to advance, one tooth at a time. The saw tooth imparts enough force to maintain the swing of the pendulum and to advance the clock's mechanism displaying the time. The force was provided by a weight which slowly descended imparting enough energy to keep it all running, and the pendulum's length can be adjusted to keep the clock accurate. Galileo did not make a working pendulum clock, nor did his son who took up the task after his death. There are many subtleties in this majestic machine which are best explored in works dedicated to this device, but it's worth emphasizing that a well-made pendulum clock could provide a very accurate measurement of time. The first such clocks appeared in 1656, invented and patented by the Dutch scientist Christian Huygens.

Originally clocks were not all that accurate and were therefore provided with only hour hands. Before the pendulum was introduced a clock might vary by 15 minutes a day. The pendulum improved the accuracy to 15 seconds a day, but they were also large and required maintenance and were expensive and – well for this reason were usually placed in towers for everyone to see and hear. For the majority of the common folk, time was not yet important. A call to church services maybe, and when industrialization took hold, a call to work in the factory. A village clock was quite sufficient. But as clocks became more accurate and time became more important this all began to change.

Early in the process the night became included into the "day" and the length was logically 24 hours with equal hour lengths. Clocks ran all the time, night and day so it was foolish not to include the night. This made clock manufacturers happy. Even if the night hours were still not of much use, they were there at least. The clock itself developed in three distinct directions.

The first was a more accurate time piece. Cost was the first obstacle. Better metallurgy and more precision in the manufacture of the parts made the clock not only more accurate, but more affordable. Mantle clocks began to appear in middle class homes regulated by pendulums and escapements but powered by springs. The most accurate pieces were the tall case clocks, and they became something of a status symbol, much like smart phones today. They were elaborately made and decorated, as much furniture as timekeepers, but they also announced that the owner was important enough to need to know the exact time, even if he didn't really need to know it. Many of these clocks are quite elaborate and a few years ago enjoyed a rebirth of appreciation. They were sought after antiques or reproductions of antiques that graced the homes of many people who enjoyed them even as they consulted their digital watches to find the time. The clock's chimes were reminiscent of the "olden days", nostalgia at its best. Incidentally, the world's most famous clock, Big Ben in

London, England, was in fact built relatively recently. It was completed in 1859 and was the largest and most accurate four-faced striking and chiming clock in the world. It is located in a tower at the north end of the Palace of Westminster that was originally named the Clock Tower, but the name was changed to the Elizabeth Tower in 2012 to honor Queen Elizabeth II's diamond jubilee. Its chiming pattern is still referred to as *Westminster Chimes*. This clock is pictured on the cover of this book.

    A word here about striking and chiming clocks. A striking clock has a sound at the hour, usually announcing the number of the hour and then a single strike at the half hour. A chiming clock has a tune or tattoo on each quarter hour with a strike for the number of hours on each hour. A striking clock generally requires two springs or weights, one for the time mechanism and one for the strike, while a chiming clock has three driving forces, one for the timing works, and one each for the chimes and the hourly strike. These are often accessed through the winding holes in the face of the clock and therefore a striking clock is often referred to as a *two-hole clock*, while a chiming clock is said to be a *three-hole clock*.

    As they progressed, the clock explored new areas as novelty pieces, the second avenue of development. I will mention two of these because they did not and were not really intended to tell time. They were ornamental time pieces. The first that I will discuss is the cuckoo clock. This still popular wall clock could be fairly accurate but was designed to entertain. Originally it cuckooed on the hour but rapidly added wheels with singers and dancers twirling out of doors in an overwhelming display. Their cases were elaborate as well. The trouble was that the power, weights usually, was too small to support the display for more than a day. That is, they had to be wound every day, while the standard clocks were seven- or eight-day machines. They still had their place and surprisingly were manufactured in Switzerland where many of the most accurate watches were also produced.

 Another clock I will mention on the other end of this spectrum is the so-called anniversary or four-hundred-day clock, or properly the torsion pendulum clock. This torsion pendulum was invented in the United States by Robert Leskie in 1793 (the same year as the French Constitution establishing the First Republic, and the beginning of Washington's second term as President of the United States). The invention was applied to a clock escapement by Aaron Crane in 1841 and again by Anton Harder in Germany in 1879. Anton was apparently inspired by the rotation of a chandelier after a servant turned it to light the candles and it kept rotating when he let it go. Chandeliers were apparently a popular thing for clock inventors to spend their time watching (see comments about Galileo above). The trouble with this clock is that it didn't work in spite of the efforts of three inventors. Eventually the Germans developed better machining techniques from which the clock benefited, but inherent in the style was the undoing: the clock was supposed to run for 400 days without any attention. Eight-day clocks were wound, or weights lifted once a week and therefore the time could be adjusted once a week. The time could be corrected, and the pendulum or suspension adjusted to ensure better accuracy. A clock left untouched for a year was unlikely to keep anything close to the correct time. Fifteen seconds a day is more than an hour and a half in a year. Fifteen seconds in seven days is less than two minutes. Further, the mechanism was sensitive to air movement (most were placed inside a glass dome) and temperature. It was beyond the existing technology to expect these pieces to keep accurate time over an entire year. They were truly more an expression of art and the fascination with things mechanical than any attempt to keep their owners informed of the accurate time. Evidence of this is that the works are left fully visible inside their glass domes.

The third line of development for the mechanical clock was one of a purely practical nature. Ships at sea needed to know where they were. The determination of latitude, the position north and south, was relatively simple. The sun at its zenith can be easily measured and if one knows the date, a table will give the latitude. Longitude was more difficult. The vessel was moving along with the sun in an east/west direction. In relatively small bodies of water, this was a "small" problem. In the Mediterranean Sea for example, the ship could sail more or less in the direction of the desired destination, for a short time usually, and run into land. "Mediterranean" means *between land* after all. The charts used were designed for this. Major and minor ports were shown and "roses", radiating lines denoting direction, were placed over these ports giving the compass direction to reach another port. If there was a little drift, no problem. The charts provided landmarks, literally markers on the land, i.e., mountains or light houses etc., to allow navigators to follow the coast to find the port they wanted. This sailing method was "coasting". I will leave the reader to explore the transferring of this term into our modern lexicon. The term meant the same then as it does now: to follow a known easy path.

Sailing on the ocean was a more significant problem. The solution was approached from two different directions. An astronomical solution fits well into the current navigational experience. Navigation was carried out by observing the stars, the sun being one of them. While there eventually was a celestial tool developed to answer the question, it proved tedious and required a great deal of skill. The obvious simple answer was to compare the skies where the vessel was with what they were at a known point. The earth rotates moving the apparent position of the heavens and if the observation is made at the same time then it becomes easy to calculate how far apart the two points are. The key word in this is: *same*. The observer on the ship must know the exact time in order to know the time difference between his vessel and a known point of reference. In 1675 Greenwich Mean Time was established as a standard universal time and the Royal Observatory was built.

We now return to the clock. Accurate time pieces were being developed, even time pieces that could maintain accuracy over several months were not out of reach, but to maintain accurate time at sea was a problem. The tall case clock would not work. The pendulum escapement was dependent on the movement being undisturbed, something no ship, even in the calmest of seas was capable of. Without exploring the mechanical and political, not to mention egotistical, adventures that helped and hindered the quest, I will jump to the end. John Harrison, an 18th century clock maker, finally developed a clock sufficiently stable and accurate to allow the easy calculation of longitude. There is a detailed account of this story in Dava Sobel's excellent book: *Longitude: The True Story of a Lone Genius Who Solved the Greatest Scientific Problem of His Time*.

The key was the balance wheel escapement. This device consists of an oscillating wheel exquisitely balanced and substituted for the usual pendulum. It is "balanced" to eliminate any effect of gravity or motion with the coiling and uncoiling of springs providing the oscillation that is necessary. This escapement is used in carriage clocks, which can be carried in carriages or on trains and other vehicles, in watches and in the chronometers carried on ships. It was the standard until the advent of the GPS system. All of these uses required a clock that was immune to the effects of gravity and motion, but only the ship's clock required a precise time over many months or even years at sea. They became indispensable in the years of exploration and lengthy sea travel. In the case of watches, it became another of the status symbols of the past age.

Meticulous naval inventories show that HMS Beagle carried a total of at least 34 recorded chronometers over the span of its three main survey voyages from 1826 to 1843, and 22 on the second voyage with Charles Darwin on board, when they had a dedicated cabin for the chronometers. Some were Navy property and others were on loan from the manufacturers, as well as six on the second voyage owned by the captain, Robert FitzRoy.

I will mention the modern time pieces, clocks, if you will allow, only briefly. They are in fact rather boring compared to their predecessors. A number of accurate devices have been used. Microwave emissions are so consistently the same that they can be used as can the decay of radioactive substances. These are curiously termed, atomic clocks. The quartz movement similarly provides a constant time and is cheap to build and remarkably resistant to abuse. The modern ship's clock is one of these. Almost all timepieces in the United States are now connected to the NIST (National Institute of Standards and Technology) central "clock" located in Boulder, Colorado. It is an atomic clock controlled by the emission from radioactive cesium, referred to as a *cesium fountain clock*. This keeps them on time and adjusts for location through the cell network's GPS and makes correction for things like location and daylight savings time. Similar systems exist around the globe.

The second was defined as a result of the cesium clock in 1967. One second was defined as "the duration of 9,192,631,770 periods of the radiation corresponding to the transition between the two hyperfine levels of the ground state of the cesium 133 atom". This refers to a cesium atom at rest at a temperature of absolute zero. The people who wrote this definition are the clear winners of the most incomprehensible definition of time ever written. Cesium is also spelled caesium, just to add to the confusion, and was not discovered until 1860. If you have nothing to do with the rest of your life you could figure out how many *"periods of the radiation ... of the cesium atom ..."* have passed since cesium was discovered. (https://astronomy.swin.edu.au/cosmos/s/Second)

I will not discuss the art of modern clocks which basically attempts to express a nostalgic or an ultramodern design. The watch has become a status piece of jewelry. I'll cite the transient fascination with digital watches and the retreat to the analog as evidence. Digital time is still dominant on electronic devices such as computers and cell phones of course, the principal time source for most people.

The people who wear these aforementioned "status" watches usually tell time from their cellphones anyway. We leave it to future generations to attempt to explain why we developed a passion for precise time but were still incapable of getting anywhere on time.

As time pieces became more accurate it became possible to divide the hour into minutes and the minutes into seconds. With the demands for precise time on ships and trains, these divisions became necessary. Note that old mathematics still dominated and 60 was the magic number, not a base 10. Before I move to the smaller and larger measures of time: millions and billions; micro and nano, I must visit two other features of modern time mentioned above: time zones and daylight savings time.

Time zones are simpler if somewhat quirky. At first there was no need for these. Local time was all that mattered. A person could only travel a few miles in a day on land and perhaps a hundred on board a ship. Time of day was determined by the sun; its zenith being the noon hour (or in more ancient times the sixth hour) referred to as the mean solar time. Time was different at every longitude, 4 minutes different for every degree of longitude and as discussed above, precise local time is necessary in order to calculate longitude. The approximation inherent in time zones interferes with this calculation. But local time sufficed until travel was sped up and time became critical.

We have discussed the need for accurate time in sea travel, but this was a minor concern compared to railroad travel. Two trains cannot pass each other on a single track as two ships can pass at sea. Train schedules had to be more precise and apply to all the trains sharing a system. In short all the trains had to operate on the same time, not the local time, in order to coordinate their schedules. This was easily done in a small country such as Great Britain, but for the United States time zones were necessary.

In 1847 the British Railroad Time was established so that all traffic could be synchronized but it was really only applied to the railroad and telegraphs which were generally run by the railroads in England. In The United States the problem was not as easily solved. Geography dominated here. The United States is a large country. At first each railroad company determined its own standard time across its entire network. This led to stations served by several companies having several "times", different for each railway line. You could literally depart on a train leaving hours before the train on which you came on arrived. This seems confusing to us and was even more confusing to people operating or riding on the rails.

In 1863 Charles Dowd proposed a system of four zones, but it went no further than the idea. He pursued this with some modifications, but the railroad companies expressed no interest until 1883 when a system proposed by William Allen was adopted. The four boundaries were determined by railroad centers and included both the United States and Canada. This system was gradually adopted for all businesses and eventually all government and social activities. Detroit was the last major city to officially adopt it in 1918. In rural areas, local time still prevailed since farming was inextricably tied to the Sun and the United States was still a rural society. The animals and fields did not follow the clock, but we are about to see this confrontation in the next paragraph: Day Light Savings Time.

The purpose of Day Light Savings Time is – well to save daylight. In the summer months, particularly in northern latitudes, the daylight becomes significantly longer than the night hours. In the far north the Sun never sets for a month or two (depending on how far north you are), and correspondingly never rises for a month or two in the winter season. At the equator the day and night are monotonously equal. In the southern hemisphere the same is true, but there are fewer people affected. Incidentally, the total yearly day time and nighttime is the same in all latitudes.

The daylight that is "saved" is really just borrowed from the early morning when urban, some would say civilized, people are still asleep. This "saved" daylight is added to the evening when people are home from work and ready to enjoy their leisure. This is accomplished as everyone knows by shifting the clock forward an hour in the spring and back in the fall. The state of Arizona still refuses to adopt this even today, citing the oppressive heat and sunshine that afflicts the state. I have seen it suggested that the conservative anti-central government nature of that state may play a role as well. Several states at the same latitude as Arizona have no difficulty with daylight savings time.

The history is a bit more understandable. Day Light Savings Time was instituted during World War I, to save fuel and electricity. It was accepted in urban areas, but scorned in the farmlands. The cows did not change their habits by the clock. When peace came, daylight saving was abandoned. With the Second World War, the change of the clocks was again instituted, but with much less resistance. The United States had shifted from a majority rural to a majority urban society between the wars.

Now I will say a few words here about the *watch*. I consider this to be a timepiece that is small enough to be worn as on a wrist, or carried in a pocket, although some were rings worn on a finger. These were relative latecomers in this story, requiring advanced metallurgy and machining techniques and therefore became common only a few hundred years ago. They have always been a synthesis of time pieces, status symbols and jewelry. The only interesting thing for the purpose of this writing is that they were regulated by the balance wheel escapement the same as nautical chronometers and carriage clocks. They are often classified by the number of jewels they have, which refers not to their embellishment as jewelry, but to the bearings within them, which were hard jewels, often diamonds, to avoid wear and subsequent failure. They were and still are often covered with precious and semi-precious stones as any piece of jewelry would be. The wristwatch was the favored retirement gift a half century ago. My father received a gold (plated) one when he retired. Of course, since he was an engineer he also received a butane cigarette lighter with a drop of oil and lump of coal visible inside the cubical glass base. It sits on my desk today, unused of course. More has changed than has stayed the same. Modern watches no longer have jewels as bearings since they have few moving parts and are so inexpensive that they can be easily replaced.

Some watches were true chronometers, however. The conductors on railroads and trolleys are often pictured looking at their pocket watches. It was, in fact, their job to keep the vehicles "on time". The engineer ran the train or trolley, but the conductor made sure it stayed on schedule. The bell on the trolley signaled to the engineer to stop or go faster or slower. Similar signals were used on the railroad trains.

Most watches today are electronic, quartz or some other extremely accurate device, and are controlled by the NIST system except for those rare pieces owned by those obsessed with antiques and less with precise time measurement (or with cost). For the rest of us, we all are connected and live in, the same time down to the fraction of a second.

The final "advance" in time came with the advance of science. Time had been concerned with days and years, minutes, and hours; but science demanded both larger and smaller measures. Two or three hundred years ago the idea of periods of time in millions of years or fraction of seconds did not exist, not just because they were impossible to measure, but because there seemed no use in them. It was a simpler time, perhaps a more honest time.

In times past, more than a couple hundred years ago that is, "science" was referred to as Natural Philosophy as I have said. It was this that Shakespeare has Hamlet reference when he says: "There are more things in heaven and earth, Horatio, than are dreamt of in your philosophy." (*Hamlet*, Act 1, Scene 5). Science was but one way of viewing the natural world and was thought of as a model, not an absolute truth. Copernicus has a disclaimer in his groundbreaking work: *On the Revolutions of the Heavenly Spheres,* stating that his ideas regarding rotation of the planets around the Sun, not the Earth, is only a more convenient and more accurate model allowing more precise calculations of their movements. These studiers of Natural Philosophy: Copernicus, Kepler, Newton et al, all realized they were only describing a more accurate theory, not providing the ultimate answer.

As science progressed, however, it needed larger and smaller measures of time that would not be bound by the older mathematics. That is, they would be measured in the base 10, not the numbers divisible by threes and fours, etc. It's worth speculating briefly here that the dominance of the base ten may have arisen because we have ten fingers on which to count. The billions of years the universe has existed or the millisecond or nanoseconds that mark the time for subatomic particles are in "scientific" notation. They are larger and smaller than anything "dreamt of in your philosophy" but should be taken with a dose of realistic modesty. They are not final answers, at least not yet, only a little better answer and one day they will fall when a new measure shows itself.

Old theories are replaced by new theories to be replaced in turn by newer ones. Only one thing remains certain: time passes whether we measure it or not.

About the Author

Paul Janson is a retired emergency medicine physician who has filled his retirement by pursuing his love of literary expression and his love of mysteries, particularly those involving the practice of medicine. He has published six novels, a non-fiction naval history and a dozen children's picture books the latter based on the adoption of his two daughters, all in just four years.

In the meantime, he lives with his wife, Mary, of 49 years and his daughter, Emma, adopted from Ecuador, and he is helping run a family-owned ice cream shop in Groveland, MA. If you're ever in Groveland drop by ***Jeff and Marias Ice Cream and Food Shop*** and have some ice cream and maybe pick up one of Paul's books too.

Paul's web site:
pauljanson.com

Paul's blog:
https://pauljanson.wordpress.com/category/philosophy/physics

Other books by Paul Janson include

Mal Practice, a mystery of medicine and murder.
Pediatrician Joe Nelson is being sued in a malpractice case involving the death of one of his patients, when he discovers that his patient's death was not an accident, it was murder. Soon he becomes the next target.

With a Little More Practice: a sequel to Mal Practice.
Joe Nelson and his friends from **Mal Practice** are in Las Vegas when murder intrudes on their vacation. The victim is a young runaway and Joe can't ignore this even when he finds himself at odds with the police.

Scratch, a young adult novel about a magical cat.
Onyx is a magical cat. He saves lives by scratching people and two young sisters are the only ones who seem to realize just how special their cat is until the events prove them right.

The Manuscript, a thriller of nuclear terrorism
A literary agent and a retired police sergeant, Joan and Frank, uncover a plot to detonate a dirty nuclear bomb when an inmate sends a query letter about a manuscript he has written. They are nearly killed trying to get someone to believe them.

Advance Directive: a sequel to The Manuscript.
Someone is killing the patients in the local nursing home to get their inheritance. Joan and Frank from The Manuscript once again have to stop the murder, but first convince someone that the deaths are murder.

The Ice Cream War, a mystery of hot fudge and murder
The story of two rival ice cream shops run by Mary and Jerry and the body found in the one run by Jerry. There is murder, humor and of course a little romance as well. The murderer may surprise you, but the romance will not.

The Child in Our Hearts.
A children's picture book about adoption and based on the belief that all children begin in the hearts of their parents regardless of how they come to be in a family and regardless of what kind of family they become a part of. There are several versions of this book for different families, including two mothers and two fathers, and single parents. Some are available in Spanish as well. There is also a version for Assisted Reproductive Technology births. A worthwhile reading for all children and maybe for all adults too.

Battles and Battleships, a narrative history of naval warfare from 1866 to 1905
Paul's first non-fiction work encompassing the development of the battleship and the lessons that can be learned and applied to our current time.

Murder Undone
A collection of short stories concerning the act of murder and the consequences explored from many points of view. There are lawyers, physicians, ghosts and even a realtor in these easily read offerings.

www.ingramcontent.com/pod-product-compliance
Lightning Source LLC
Chambersburg PA
CBHW021932170526
45157CB00005B/2296